S0-BOC-775

Father Mouse had a lovely daughter.
Many a mouse wished to marry her.
They bought small gifts for her. They
also brought sweets and treats.
Sometimes they even fought over her.

But her father would only laugh.
"Marry *only* a small mouse? What
kind of talk is this?" he said. "My
daughter won't fall so low! Her husband
must be very great!"

"She should marry the sun!" he thought. "He always shines and warms the world. Sun is the greatest of all! I will go and call on him."

So, he set off to walk up the mountain.

"There is one greater," Sun told
Father Mouse. "Cloud is far greater
than I. It is Cloud who always shades
my shine!"

"I must talk to *him*," thought Father
Mouse. But when he did, Cloud let out
a laugh.

"Why, Wind is greater than I!"
Cloud said. "He can blow me to
small bits."

Now Father Mouse sought the wind.
But Wind sent him off also by saying,
"It is Mountain you seek."

"Mountain is greater than I," Wind told him. "Only *he* can stand still when I blow."

Father Mouse had nowhere to walk. He was *on* the mountaintop. And Mountain had heard all the talk.

Mountain said, "It is not I you ought to call on. For a small mouse is greater than I. Only *he* can fill me with holes. *His* digging crumbles my walls! Find a *mouse* to marry your daughter!"

And *that* is just what Father Mouse
did! For he had found the truth:
A small mouse is the greatest of all!
(And Mouse's daughter thought so too!)